Waves and Rays

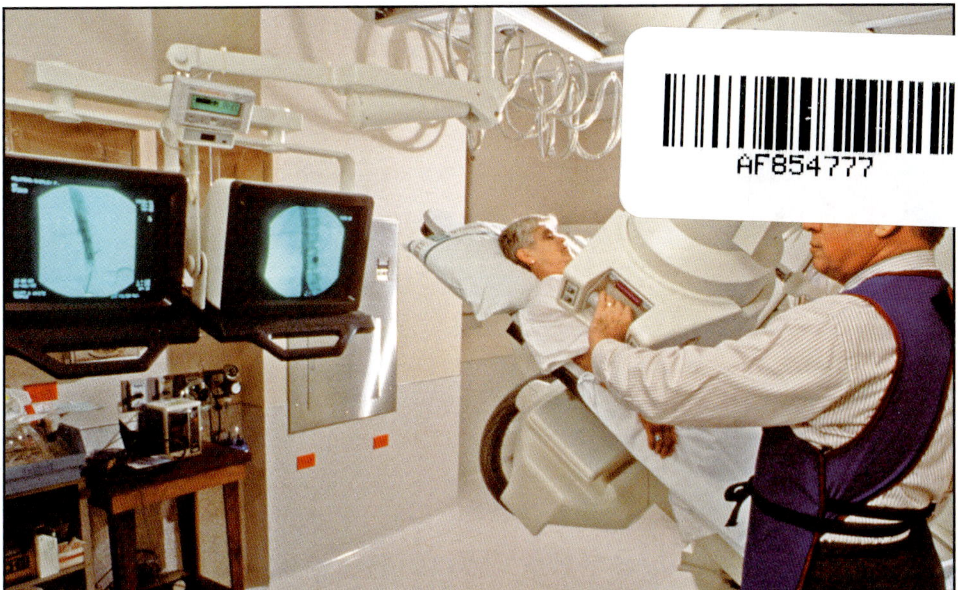

by Paul A. Barra

HOUGHTON MIFFLIN BOSTON

Water waves are among the few waves that humans can actually see.

Waves and Rays

Imagine being able to see all the waves that exist in our environment. If sound waves, brain waves, heat and radio waves were visible to you, you'd need help to find your way to the refrigerator. Everything would become blurred by waves, and you'd have trouble seeing your friends' faces. Fortunately, all of those waves are invisible, and so are most others.

You can see waves in the ocean and other large bodies of water. The only other waves you can see are those of the visible light spectrum. These light waves come from the sun or from electric lights. There are many types of electric lights, including mercury vapor lamps, neon tubes, and electronic LEDs. The two in homes and schools are either incandescent or fluorescent. Incandescent lights are the common bulbs that go into our lamps. They consist of a thin wire in a bulb of gas. The wire glows when heated by electricity. The bulb also gets hot.

Fluorescent bulbs burn cool and last much longer than incandescent ones. They consist of a sealed tube coated with a substance that produces light just like sunlight (that is to say, visible light).

Visible light reflects all the colors of the rainbow. It starts at red and goes through orange, yellow, green, blue, indigo, and violet (ROY G BIV). Seen all together, visible light looks white to us. If white light shines on a red apple, say, then only the red band is reflected and you see the color red on the fruit skin. You see black when no color is reflected. For instance, if you use a green bulb to light up a room, you can only see color from the objects that can reflect green. Since it has no green on it, the red apple will show up as black.

Colors are important to people. Some will choose a car to buy because they like the color, even if a different one might be less expensive or get better gas mileage.

The Electromagnetic Spectrum

This visible light spectrum is part of the electromagnetic spectrum (EMS). The EMS is a wide band showing all types of radiation. As shown here, their wavelengths can be long like radio waves, short like gamma rays, or in-between like the visible waves. Waves are sometimes called rays and sometimes they are called radiation. Whatever they are called, all the waves of the electromagnetic spectrum travel at the speed of

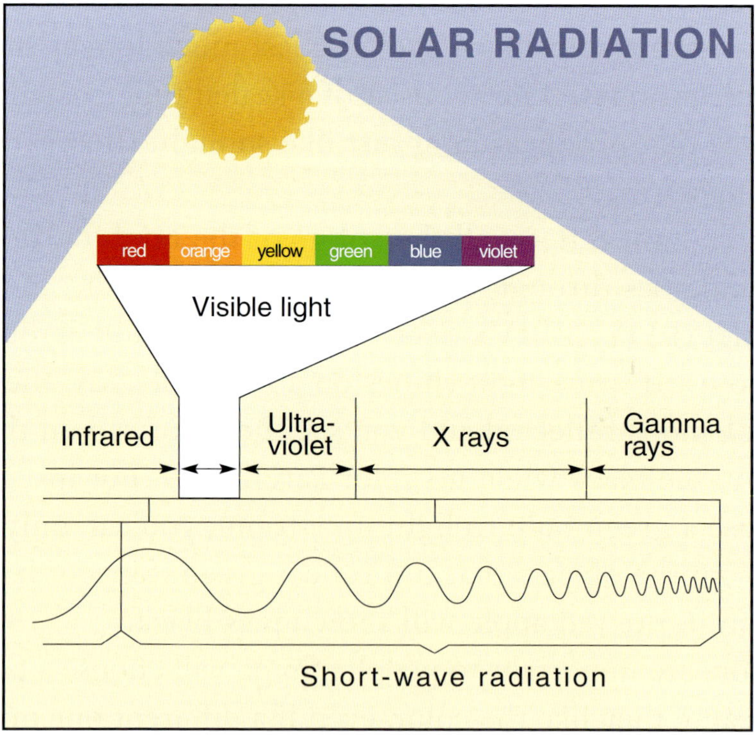

The electromagnetic spectrum

light (precisely 299,792,458 meters per second, or about 186,000 miles per second). That's fast! In fact, nothing in the known universe can exceed the speed of light.

All the waves in the EMS travel at the speed of light, even the ones you can't see, and they all have different levels of energy. That includes visible light, the type of light you are able to see.

What else is in the EMS? Well, there are the waves just to either side of the visible light spectrum, infrared and ultraviolet. Infrared radiation is next to the red colors you can see (*infra* means *below*).

Infrared Radiation

Although you can't actually see infrared radiation, you can feel its heat. All warm objects give off infrared radiation.

Rattlesnakes and some other reptiles have sensors on the sides of their faces that detect infrared radiation. Small animals are warm, so these snakes can use their sensors to track down prey. Snakes don't see well unless an object is moving, so heat sensors could be valuable in locating something to eat before they can see it.

People don't have a sensor to detect infrared radiation, but we do have a way of detecting French fries being warmed under an infrared radiation lamp. If you put your hand under one, you'll feel the heat. Many restaurants use them. They keep food warm and they're safe.

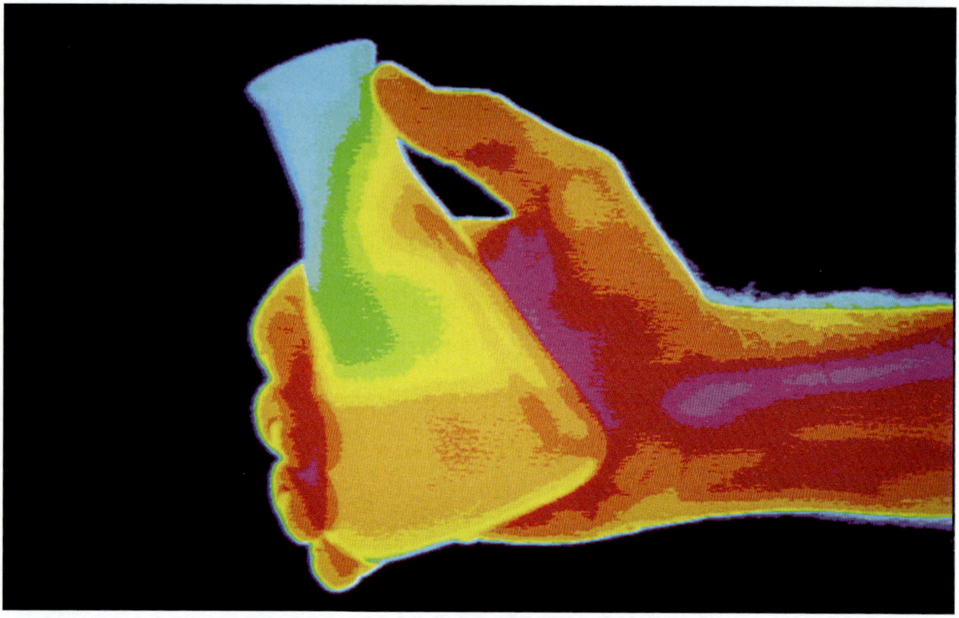

An infrared photo shows heat; yellow and red are the warmest.

Other businesses use scanners to pick up differences in the intensity, or strength, of infrared waves. For example, doctors can scan a person to locate tumors that cause hot spots in the body. Engineers can scan a building to find out where it's leaking heat. Astronomers can even detect warm dust in space from new stars that are not hot enough yet to give off visible light. The Army uses infrared scanners to detect enemy soldiers at night. If you were a spy using a night scope, you could then switch to your camera with special infrared film and shoot pictures of suspicious people. They'd never know you were there.

Even the remote for your TV works by shooting a beam of infrared radiation. You can't see the beam the remote gives off, because it's outside of the visible light spectrum, but the television sensors react to it. Try changing channels when someone is standing between your remote and the television. It won't work.

Burglar detectors shoot a beam across a doorway into another kind of sensor. If you rushed into a friend's house and the burglar alarm included a beam and sensor, you would pass through the beam and block it from the sensor. You wouldn't get hurt, but the alarm would sound.

Ultraviolet Rays

Ultraviolet (UV) light starts at the opposite side of the visible light spectrum from those infrared beams, beyond the violet end of the light we can see (*ultra* means *beyond*). UV light comes mainly from the sun and is filtered through a type of oxygen in the Earth's atmosphere known as ozone.

Welding torches and a few other inventions can generate UV radiation. The Sun produces most of it, however, and you may have heard about the damage it can do to you. UV-B radiation is the medium-length ultraviolet wave. It's the dangerous kind. Long-term exposure to the UV-B rays can cause skin cancer. Reflected UV rays can cause snow blindness. Medical professionals tell you to protect your skin with clothing or sunscreen and to wear protective sunglasses when you're out in the sunlight. Your eyes are particularly at risk to UV light. Ultraviolet exposure can make your eyes extra-sensitive to light and make you feel as though you have sand in them. Welders must wear special goggles to protect their eyes from the UV light their machines produce.

Not all UV exposure is harmful. Some is actually good for you. It stimulates the production of vitamin D, which our bodies need. UV light can be used to help some skin conditions, as well.

X-rays

X-rays are the next radiation to ultraviolet. They do not occur naturally on Earth but are made by machines. Because they can penetrate the soft tissues of our bodies and reflect off hard parts, X-rays are helpful to doctors and dentists. They use them to get a look at our bones or the parts of the teeth beneath the gums. X-rays are created by powerful electricity, which causes electrons to jump off a metal in a tube. The radiation produced is high energy and dangerous. In fact, Thomas Edison's assistant died in 1904 from cancer caused by X-rays. He used to X-ray his hands to demonstrate to people the penetrating power of the radiation. Many other X-ray pioneers suffered the effects of radiation poisoning too. Today, people who work with X-ray machines wear protective gear or stand behind a lead barrier when using the machines.

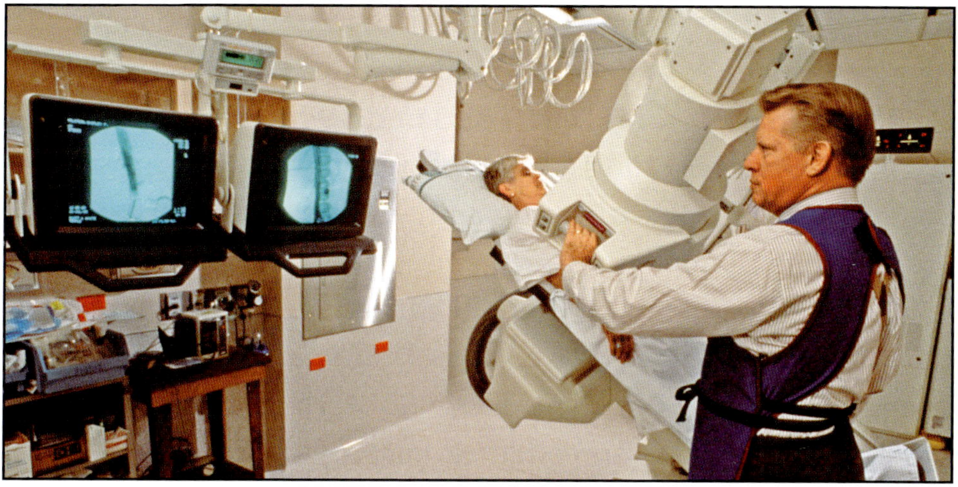

An X-ray is a form of electromagnetic energy made by machines.

Hot gases in space also produce X-rays, but they are not harmful to people on Earth.

The X-ray machines in airports are not harmful to you either. Security guards may X-ray your luggage, but not you. A new approach to people screening is a device that gives off what are known as backscatter X-rays. These are low power rays that see through clothing. The federal government says that these new devices are not dangerous at all.

Gamma Rays

Gamma rays are the shortest waves and are emitted by certain stars and other celestial bodies. These waves are so powerful that food scientists use them to kill microbes on fruit and meat. It's a process called irradiation. The gamma rays zap the microbes and then disappear. When you eat the food, you're not eating germs or gamma rays. Gamma rays are used in other settings too. Hospital workers sterilize medical equipment with gamma rays. They use special lead-lined containers because gamma rays cannot penetrate lead. They can penetrate and kill germs, though. Gamma rays are also used to kill cancer cells, but they would be equally dangerous to your healthy cells if you were exposed to them. Nuclear technicians target cancerous growths with narrow beams of gamma radiation, trying to kill only cancer cells.

Gamma rays are emitted from radioactive elements on Earth. A radioactive element is one that decays naturally into another element, giving off radiation from its nucleus as it does so. Gamma rays are powerful enough to pass through concrete a meter thick. Radioactive elements, such as some types of uranium, plutonium, and radium, occur naturally.

Nuclear power plants produce other gamma rays. Nuclear plants are designed to contain the dangerous gamma radiation. If you worked at one of these places, you'd have to protect yourself by staying outside the place where the nuclear fuel is kept. You'd have to handle radioactive fuel pellets by maneuvering robotic arms from outside the container vessel. Mechanical hands on the arms grasp the pellets.

Madame Curie discovered much of what we know about radioactivity.

Madame Curie

Marie Sklodowska Curie figured out much of what we now know about radioactivity. Marie Curie was the first person to ever win two Nobel prizes, one in physics and one in chemistry. She died in 1934 from radiation poisoning. At one point, Madame Curie had an entire chemical company working on tons of pitchblende, the ore that contains uranium. Just as with early X-rays, there's no telling how many people were affected by gamma rays before anyone knew enough to protect workers.

Microwaves

On the other end of the electromagnetic spectrum from X-rays and gamma rays are microwaves. They are not dangerous like gamma rays or X-rays. Microwaves are slightly longer in wavelength than infrared radiation. They are emitted by exploding galaxies called quasars and by clouds of gases in deep space. We can pick up these waves on radio telescopes and "see" what is going on out there. Since microwaves also travel at the speed of light, we can get up-to-date information that way. Of course, the distances in the cosmos are so immense that the term up-to-date must be considered relative. After all, it takes a full eight minutes for light from the sun to reach us. And the sun is the nearest star to Earth.

Perhaps engineers didn't want to wait that long for the microwave signal, so they learned to make them. They are used primarily for line-of-sight communications. Those are the kind that travel straight ahead and do not bounce into the atmosphere or bend around mountains. They travel a maximum of about 50 kilometers (30 miles). Some are so short they are measured in centimeters, so they got the name microwaves. They are longer than all light waves except other radio waves, however.

Microwaves can pierce fog. These waves are relatively benign (harmless) when they are received by a radio. They are the active component in radar ("Radio Detection And Ranging"). Radar transmitters send short bursts of microwaves. Objects that are hit by the waves send back echoes. Electronic devices record the echoes and display them on a screen. The closer an object is, the stronger is its echo.

Satellites in space and spaceships also generate microwaves. They record the objects below them on the surface of our planet. With computer enhancement, microwaves show ice floes in northern seas and enemy tanks on the ground.

Wireless cell phones are actually two-way radios that send and receive converted microwaves. The Federal Communications Commission (FCC) regulates the strength of microwave signals used for communication. The amount of energy cell phones are allowed to use is so low that cell phone companies divide up their territory into cells. That's why wireless phones are called cell phones. Towers or base stations relay signals. Some people worry about the effect of having microwaves near our heads on a regular basis, so the Food and Drug Administration (FDA) tests the safety of wireless phones. So far, the FDA sees no danger in cell phone use.

For now, one of the most popular uses of microwaves remains the oven that bears its name. An inventor named Percy Spencer developed the microwave oven in 1946, when he was trying to improve radar. A Magnetron tube produced longer microwaves than the ones being used in radar. When he was standing near one, the microwaves excited water molecules in a chocolate bar in his pocket—and melted it. Spencer realized that the waves were hitting the candy bar. That made the water, and the food, heat up.

To be sure he was correct in his thinking, Spencer put some corn near the Magnetron. Sure enough, it started jumping around and popping. Then he beamed the long microwaves at an egg. The egg exploded.

Since practically all food has some water in it, even popcorn, microwave ovens can heat and cook food by heating the water in it. The water in popping corn is vaporized by the heat and the kernels puff up with air.

Radio Waves

Other waves at the long end of the spectrum are simply called radio waves. These are the ones with wavelengths greater than 30 cm. Special techniques impress data on the waves so we hear sounds like speech and music or see pictures. A radio wave is transmitted by a metal wire or rod. These transmitted waves are modulated, or varied, by the station sending out the signal. That's why your radio has an FM (Frequency Modulation) dial and an AM (Amplitude Modulation) one.

Television stations transmit varying radio waves just like radio stations. It's called broadcasting. These waves are converted into pictures. The signal is sent out by powerful transmitters and is decoded when a receiver, your TV set, picks up the signal.

Cable TV companies transmit waves through fiber optic cables. These are non-metallic cable tubes that can transmit light. Satellite TV companies use signals that are bounced off satellites in space and sent to your TV set.

Other Rays

The energy from the sun is converted to chemical energy in plants during a process called photosynthesis (*photo* means *light*; to *synthesize* means to *make*). Solar energy can also heat houses and power cars. The use of solar power would reduce America's use of oil, a resource that cannot be renewed. Energy from the sun, the electromagnetic spectrum, is renewed constantly. If we can develop solar power as a practical energy source, we will never run out of usable energy. It's not impossible, either. The Aurora, a totally solar-powered car, was driven at an average speed of 86.2 kph (53.56 mph).

A solar-powered vehicle

Of course, that was in a part of Australia where there is abundant sunshine. Solar powered cars in Florida or Hawaii would have plenty of sunlight available also. If you lived in a place where rain or snow was a frequent visitor, you would need to have a way to store solar power. The technology for that is getting better every year. Soon, you could be riding to school in a sun car—without having to worry about filling up the tank.